The Shergold safety bicycle, 1876-8, is believed to be the earliest surviving bicycle with a chain drive to the rear wheel. Although a crude and cumbersome machine, it marks the beginning of the safety bicycle. The basically one-piece frame harks back to the original Macmillan machine. The cumbersome pivot and rod steering coupled to the front wheel enabled it to be set well forward clear of the rider's feet and the pedals. The height of the saddle can be slightly adjusted. The primitive oil lamp is believed to be an original fitting and may have been an adaption of a penny farthing hub lamp.

THE SAFETY BICYCLE

Ian K. Jones

Shire Publications Ltd

CONTENTS

Published by Shire Publications Ltd. PO Box 883, Oxford, OX1 9PL, UK. PO Box 3985, New York, NY 10185-3985, USA. Email: shire@shirebooks.co.uk www.shirebooks.co.uk

© 1986 Ian K. Jones. First published 1986. Transferred to digital print on demand 2015. Shire Library 174. ISBN-13: 978 0 85263 804 0.

Printed and bound in Great Britain.

A CIP catalogue record for this book is available from the British Library.

ACKNOWLEDGEMENTS
Particular thanks are due to Mr John Collins, Curator of the Mark Hall Cycle Museum, for the large amount of time spent uncovering information about the machines described here. Photographs on the following pages are acknowledged to: Glasgow Museums and Art Galleries, page 5 (upper); the Trustees of the Science Museum, London, pages 1, 3, 4, 5 (lower), 6, 7. All other photographs are from the Mark Hall Cycle Museum.

COVER: *Royal Enfield advertisement. Robert Opie Collection.*

BELOW: *The Whippet spring frame, 1885, Linley and Biggs, Clerkenwell Road, London. This was one of the many early experimental designs and was popular with both riders and racers because of the extra comfort given by its ingenious sprung frame. The relative positions of handlebars, saddle and pedals were fixed by being mounted on a rigid frame which was isolated from the main frame, on which the wheels were mounted, by two moving shackles and a strong coil spring. The invention of the pneumatic tyre, a far more efficient means of absorbing road shocks, ended experiments in this direction until the work of Alex Moulton in the late 1950s.*

A McCammon bicycle, 1884. With its single tube drop frame it would have been suitable for use by a lady. It is not known if this was the designer's intention or if he was simply trying for a lighter and less cumbersome method of mounting the chainwheel and pedals. The mudguards and coil spring saddle are some of the many developments that were to add to the comfort and safety of cycling. The machine was also one of the first to have direct rather than coupled steering.

INTRODUCTION

The bicycle in all its various forms is the fastest and most efficient form of man-powered transport ever produced. This book covers the development of one particular type, the two-wheeled machine using pedals and chain to drive the rear wheel, generally referred to as the safety.

Although the first mechanically propelled bicycle of which there is any definite knowledge used a clumsy rear-wheel drive, the idea was almost forgotten for decades until technology provided a simple, effective solution to the transmission problem in the form of a chain. By 1885, the chain-driven bicycle as we know it today had evolved and developed rapidly into a bewildering variety of types. This experimentation with different designs and materials still continues.

Most of the machines featured here come from the Mark Hall Cycle Museum in Harlow, Essex, which is based on the fine collection built up by John Collins of Harlow in the years after 1948. For eighty-three years from 1896, his family were cycle dealers in Harlow and some of the machines featured were sold to local people by John's father and grandfather. The Collins collection was sold to Harlow District Council in 1978, and in 1980 work started to convert the stable block of the Mark Hall Estate into Europe's first museum devoted only to the history of the bicycle. The museum opened in 1982 and won a Civic Trust award in the same year.

A replica of the Macmillan bicycle (1839-40) made by Thomas McCall about 1860. As far as is known the original Macmillan machine no longer exists but this version built by a man who may well have known Macmillan is thought to be an accurate copy. The backbone, saddle mount and driving rods are of a far better design than that employed on the Dalzell machine of 1845. Steering would be the worst problem with this machine as the position of the pedal arms seriously reduces the distance the front wheel can be turned.

ORIGINS AND DEVELOPMENT

The earliest bicycle, which first appeared in Paris in 1791, was driven by the rider pushing on the ground with his feet. An improved version called the hobby-horse was produced in London from 1818 by Denis Johnson and numbers of them remained in use until about 1830. The weight of the machine, the great effort needed to move it and poor road surfaces caused its disappearance.

Inventors all over Europe experimented with different ways of driving two, three and four-wheeled versions of these machines. The credit for inventing the first practical pedal-driven two-wheeled bicycle belongs to Kirkpatrick Macmillan, the village blacksmith at Courthill near Dumfries in Scotland. He discovered that a machine with two wheels in line could be balanced by its rider while being driven by a system of pedals and cranks to the back wheel.

Although the use of pedals to drive the rear wheel became the hallmark of the later safety bicycle, developments from the 1860s concentrated on pedals attached direct to the front axle because this was a much simpler system.

Macmillan developed his machine between 1839 and 1840 while working in the blacksmith's shop on the Drumlanrig estate of the Duke of Buccleuch. He and his assistant first built and rode copies of a hobby-horse belonging to a Mr Chateris. Macmillan then re-designed the machine with improved handlebar front-wheel steering and powered it by two pedals attached by connecting rods to crank arms on the rear wheel. The bicycle worked well and the inventor used it several times to visit Glasgow, 40 miles (64 km) away. He took out no patents and other Scottish craftsmen copied and modified his design.

4

ABOVE: *Gavin Dalzell built this copy of Macmillan's bicycle in 1845. As Macmillan's machine does not survive it is impossible to tell if alterations were made to the original design, though this machine is better in some ways and worse in others than McCall's later copy of the Macmillan bicycle. Dalzell's machine has no brake and the backbone is in two pieces with the joint a potential source of weakness. The pedal cranks are angled to prevent them fouling the front wheel when the machine turns, and this angle can be adjusted to suit the rider's legs. The coat guard is ribbed for lightness and there is no saddle spring. The connecting rods are incorrectly aligned. Copyright of Glasgow Museums and Art Galleries.*

BELOW: *A Lawson and Likeman bicycle, 1876, Lawson's second experimental machine. It closely resembles the Macmillan type and seems to be an attempt to improve the design and overcome the lack of an efficient chain to drive the rear wheel. The pedal arms are pivoted well clear of the small steering wheel. The rear driving wheel has been made as large as possible to give greater speed and the rider is seated low down for greater stability. The bar on top of the front wheel has footrests at its tip, which enabled the rider to lift his feet off the rapidly moving pedals when travelling downhill.*

Amongst these craftsmen was Gavin Dalzell, a cooper of Lesmahagow, Lanarkshire, who built his version in 1845. This machine survives and is the oldest pedal-driven bicycle known. In the following year he beat the Royal Mail coach by a handsome margin for a wager, riding round it three times. Other copies were produced including one by Thomas McCall about 1860. He was a joiner and wheelwright in Kilmarnock and had been brought up near to where Macmillan lived. Surprisingly no attempt was made to develop this machine, partly because of the popularity of the front-driven boneshakers and ordinaries (or penny farthings as they were nicknamed) and the lack of an efficient means to drive the rear wheel until the advent of the chain in the 1870s.

During the late 1860s and early 1870s some inventors worked on safer alternatives to the penny farthing. Designs by Thomas Wiseman and Frederick Shearing published in the *English Mechanic* in 1869 are generally accepted as the first safety bicycles. Shearing's machine was the most advanced, with front-wheel steering and the back wheel turned by a belt driven by pedals mounted between the wheels. This machine was not produced commercially, but in 1870 the firm of Peyton and Peyton in Birmingham produced a machine based on Macmillan's design, though only a few seem to have been sold.

The first practical safety bicycle which is certainly known to have been built and ridden was produced by H. J. Lawson in 1873-4. The machine had two wheels of equal diameter with the rear wheel driven by a chain. Lawson used this machine frequently in Brighton and the surrounding countryside. In 1876, he produced a second, less advanced design with a small front wheel and large back wheel driven by a series of cranks and connecting rods similar to Macmillan's system. During the same period a Gloucester shoemaker, Thomas Shergold, was experimenting with the safety bicycle. Sometime between 1876 and 1878 he produced the earliest chain-driven safety to survive. He rode the

The Lawson Bicyclette, 1879. Probably made by the Tangent and Coventry Tricycle Company. The first efficient chain-driven safety bicycle, its frame resembled the later cross-frame design with its one-piece backbone incorporating a vertical support forked at the base to hold the chainwheel and pedal sprocket. Because the saddle was set towards the rear of the machine the handlebars had to be connected to the steering head by a coupling rod. The front wheel is unnecessarily large, an idea derived from the penny farthing. As the machine has a fixed wheel drive footrests are fitted on the front fork. The saddle has a wide range of adjustment. Neither the surviving pedal nor the chain are original.

An 1884 Humber, made by Humber and Company of Beeston, Nottinghamshire. This machine is thought to be the first to abandon the single backbone tube design and adopt the more compact and stiffer diamond-frame form which later became universal. This early type lacked only one major feature, the central vertical seat pillar tube. In order to help the rider get on the machine while pushing it a step is fitted on the end of the rear axle mount. As the freewheel had yet to be invented rear wheel and pedals would always turn together, making mounting hazardous. Footrests are mounted at the base of the steering column, which also has a coil spring suspension system, probably the first to be fitted to a bicycle.

machine as far as Birmingham and offered the design to the Starley Brothers, Britain's major cycle manufacturers, in 1880. They refused it and, failing to raise the necessary finance, Shergold was unable to develop it and died a disappointed man in 1903.

Meanwhile Henry Lawson continued his experiments. He was manager of the Tangent and Coventry Tricycle Company and on 30th September 1879 he took out a patent for his improved safety, the Bicyclette, soon nicknamed the 'Crocodile'. At the Stanley Show of 1880, the major event of the bicycle trade year, it attracted a lot of comment. The machine eventually went into production although only a few seem to have been made. The Birmingham Small Arms Company (BSA) made two prototypes for Lawson and also designed their own version in 1884 of which they sold some 1500 machines. During the same year several

other designs inspired by Lawson's appeared, including J. McCammon's machine with a single tube drop frame, making it in effect the first ladies' safety. The Humber safety, also of 1884, was the first machine to use the diamond frame instead of the single main backbone which dated back to the days of the hobby-horse.

1884 also saw the production by J. K. Starley and William Sutton of the first Rover safety, which attracted a great deal of attention at the 1885 Stanley Show. Their second design was improved on the advice of Stephen Golder, the competition cyclist, by being fitted with direct instead of coupled steering and an adjustable saddle. On 26th September 1885, a Rover raced by George Smith set a new one hundred mile (161 km) record of seven hours, five minutes and sixteen seconds and proved the superiority of the rear-driven safety over the front-driven

TOP: *This 1885 Rover, developed by J. K. Starley at the St John's Works, Coventry, was 'the bicycle that set the fashion to the world'. Although the frame was constructed with a curved top and bottom tube, with no tube between the saddle and crank bracket, it was basically a diamond frame. At first the wheels had solid tyres but in the years after 1890 later models were fitted with John Boyd Dunlop's pneumatic tyres. These tyres and equal sized wheels made cycling safe and comfortable for both sexes and enormously popular. This example is fitted with a Miller 'Bell Rock' oil lamp.*

CENTRE: *A Swift cross-frame, 1885. The main feature of this machine is the simplicity of the frame, which consists of a backbone joining the rear hub to the steering head, crossed at right angles by a tube carrying the saddle and crank bracket. This machine was bought in 1960 from a cycle shop in Ipswich where it had spent many years fixed to the wall outside. Being in very bad condition, it needed considerable restoration. It is fitted with a Powell and Hanmer 'Oracle' candle lamp.*

BOTTOM: *A Raleigh cross-frame, 1906. One of the first of this type of cross-framed machines to be produced, it was copied by many other manufacturers, while Raleigh continued to produce it until the 1930s. The unique frame, completely different from the 1885 Swift, was designed in 1896 by a brilliant engineer, G. P. Mills, employed at Raleigh's Nottingham works. The design proved very successful and two years later a tandem version was produced. This bicycle is fitted with a Lucas 'Calcia King' acetylene lamp.*

TOP: *On the 1886 Singer, made at the company's Challenge Works, Coventry, the frame design was improved by fitting a rod between the steering head and the frame above the crank bracket. Instead of the main frame extending back as a fork to the rear hub, separate seat stays were fitted from the top of the frame to the rear hub. The steering is the pivot type and the crank bracket is adjustable for tightening the chain. The wheels have very narrow solid tyres and a simple chainguard is provided. It is fitted with a Lucas 'King of the Road' oil lamp.*

CENTRE: *The Marston Sunbeam, 1896, made by John Marston and Company Limited, of Wolverhampton, was hand built to the best standards of design, materials, manufacture and finish. It had the highest reputation for quality, durability and easy running of the few de luxe models of the last decade of the nineteenth century and the first decade of the twentieth century. This was probably the first machine to have the fully enclosed Harrison Carter chaincase, which remained a permanent feature of these models until production ended in the 1950s. It is fitted with a 'Perfecta Simplex' acetylene lamp.*

BOTTOM: *This 1889 Centaur, produced by the Centaur Bicycle Company of West Orchard, Coventry, has one of the earliest examples of the diamond frame, which remains basically unchanged today. The steering head is the early pivot type and the chain stays have an adjustment for tightening the chain. This machine was originally fitted with narrow solid tyres and later modified to take pneumatic tyres. When it was restored in 1965 they were badly decayed and had to be replaced with solid tyres as no suitable pneumatics were available, though the original valves were left in place. It carries a Brown Brothers oil lamp.*

penny farthing. In the same year the third Rover design went into production and signalled the beginning of the end for the penny farthing, whose makers responded with an immense range of modified versions, none of which would match the speed of the Rover and its companions. Amongst these early safeties was the cross frame safety popular between 1886 and the early 1900s. Many records were broken by these machines but they lacked the strength of the diamond-framed Rover. Despite the advantages of these new machines they did not seriously challenge the penny farthing until after 1890.

Solid tyres and bad roads made the early safeties uncomfortable to ride and various experiments were carried out to improve the comfort of the ride. One of the most ingenious and effective was the Whippet of 1885, but the major development was the Dunlop pneumatic tyre, increasingly used after 1890. During the

ABOVE: *A Rudge bicycle of 1889. The frame resembles the modern diamond frame but has a unique feature, the use of halved steel tubes for lightness. However, because the halved tubes are joined with metal cross pieces, they are slightly heavier than the standard hollow tube. The wheels have solid tyres and the steering is the early pivot type. The oil lamp is a Salisbury 'Invincible'.*
BELOW: *An 1896 Elswick. The frame is a robust version of the straight-tubed diamond frame with one unusual feature. To give greater strength and rigidity twin crossed-over down tubes are used. The roller lever brake to the front tyre and the saddle are probably original. The fixed rear wheel and semi-drop handlebars suggest this machine may have originally been built for racing as the freewheel was already in existence.*

TOP: *The 1905 Rover, built at the Meteor Cycle Works, Coventry, was developed from the Rover of 1885. This machine has a very large 28 inch (710 mm) example of the now standard straight-tubed diamond frame made from thin steel tubes brazed into jointing lugs. The wheels are 28 inches (710 mm) in diameter and the bicycle is fitted with a Lucas 'Silver King' oil lamp.*

CENTRE: *An Elswick ladies' bicycle of 1892, produced in Newcastle upon Tyne. This early example of a ladies' bicycle had an open frame to allow for the long dress of the period. As pneumatic tyres were still unreliable, and when punctured difficult to repair, many bicycles were manufactured with either the pneumatic or the cushion tyre. This machine is equipped with the latter to spare its rider the inconvenience of punctures and is fitted with a Joseph Lucas number 59 'Captain' oil lamp.*

BOTTOM: *A Singer ladies' bicycle of 1893, an early example of a ladies' safety bicycle built round a single-loop tube frame. A high-quality machine, it is fitted with a fabric gearcase fully enclosing the chain drive and dress cords strung from the rear mudguard to the hub. These prevented the rider's dress from being entangled in the moving parts. The machine is also fitted with a steering lock, to prevent it falling over when placed against a wall, and a Brown Brothers 'Halycon' oil lamp. The only problem a lady would experience with this machine was caused by the fixed wheel; when going downhill she would have to place her feet on the footrests on the front forks.*

TOP: *An 1894 Swift tandem. This is an early example of a racing tandem with one of the first double diamond frames, developed from the popular safety of the time with the typical upward-sloping top frame tube at the front. The front brake is the usual plunger type applying to the top of the tyre of the front wheel. The machine has a 30 inch (762 mm) front wheel and a 28 inch (710 mm) rear wheel.*

UPPER CENTRE: *A Royal Enfield tandem of 1896. This was one of the first machines made by this company. The unique frame construction has curved top tubes with the rear one lower than the front. This was one of the first 'Lady Back' tandems with what became the standard seating arrangement of gentleman in front and lady behind. It is fitted with a Lucas 'Calcia King' acetylene lamp.*

LOWER CENTRE: *A Raleigh tandem of 1898. By this time the tandem was a very popular machine. This one has an excellent frame design consisting of a single tube extending from the steering head to the back wheel, where it forks, together with additional frame tubes, which give a very rigid construction. This machine was designed by G. P. Mills and carries a Lucas 'Silver King' oil lamp.*

BOTTOM: *A Rudge-Whitworth tandem, 1905, built in Coventry. A late example of the early tandems built for the lady to ride at the front, it has coupled steering that can be operated from front and rear. The back seat is higher and the wider handlebars suggest it was intended to be steered from there. For comfort and safety the positions were generally reversed before this model was produced. It is fitted with a Lucas 'Holophote' oil lamp.*

next few years the various experimental types disappeared and the basic diamond frame machine became the accepted type. The Humber safety of 1890 had all the essential features on which the majority of bicycles have been based since. The simple diamond-frame design led to a reduction in the weight of machines to between 18 and 40 pounds (8.2-18.1 kg), and by 1895 the horizontal top tube for gentlemen and downcurved loop tube for ladies were standard for roadsters. The late 1890s saw the first great cycle boom when the rider on two wheels was the fastest traveller on the road until the arrival of the motor car. The period also witnessed the development of the free wheel and the roller lever rim-brake and the production of the first reliable gears in addition to the universal adoption of the pneumatic tyre and the roller driving chain.

As the safety bicycle developed so did the variations designed for two or more riders. The tandem proved to be the most practical and long lasting. Machines like the Rudge-Whitworth of 1905, with the lady in front and capable of being steered from both positions, soon disappeared to be replaced by the standard form with the gentleman in front and fixed rear handlebars.

WORK AND PLAY

The first half of the twentieth century saw remarkably few developments in the general form of the bicycle. Most manufacturers concentrated on producing reasonably cheap, reliable machines with the main changes being in the greater sophistication of the accessories fitted. By the early twentieth century the acetylene lamp, typified by the Lucas Acetylator Calcia King of 1907, was beginning to replace the wide selection of oil lamps then in use. Both acetylene and oil lamps remained in production until the late 1930s. By 1910, they had been joined by the first practical battery lamp and by the 1930s the dynamo was available to provide the power but the electric lamp only became supreme after the Second World War.

The firm of John Marston, however, concentrated on improving the quality of their machines and adopted the highest standards in materials and finish. The finest of these luxury machines was the Sunbeam, first produced in 1887. From 1900 it was called the 'Golden Sunbeam' and was later often known as the 'Rolls Royce' of bicycles because of its continuing superb quality. The machine remained in production until 1957 and was always in the forefront of technical advance. From the early twentieth century the bicycle had the famous oil-bath gear case allowing the chain to be constantly lubricated and in the later years a means was provided for removing the rear wheel for repairs without touching the chain.

From the earliest days of the bicycle great controversy had existed over the question of women cyclists. In the late 1880s the controversy became more acute when the development of the safety bicycle finally produced a machine that women could ride easily. Many people even questioned whether women should be allowed to ride at all and argument raged over what they should wear, when, how and with whom they should ride and the possible disastrous effects on their complexions, hair, health, families, morals and reputations. As the numbers of ladies' machines increased, so did the riders and the criticism by both men and older women. Quite suddenly in the spring of 1895 the criticism was silenced when upper class ladies discovered the bicycle. The high society cycling craze lasted barely a year but it gave women's cycling respectability and stimulated the production and development of a wide range of light, cheap machines. Gradually the barriers of convention crumbled and by the 1930s even shorts were accepted without complaint.

The tandem remained popular, particularly with married couples, in the years before motor cars became cheaper and more available. It could also be adapted for carrying children with either an additional seat mounted behind the rear saddle or a fully enclosed sidecar, or both. Many such machines survived in regular use until the 1950s.

Bicycles had been raced since the

TOP: *A BSA bicycle of 1907. The BSA company produced bicycle fittings at such a competitive price in this period that many cycle dealers who had previously built their own frames used these ready-made fittings to assemble complete machines for their customers. This bicycle was assembled by Mr William Collins, cycle dealer of Harlow, and was sold to Mr Fred Clark, the local blacksmith, for £10 10s. The machine is a typical roadster of the period with semi-drop handlebars, rod-operated caliper rim-brakes and mudguards with a reflector on the rear one. It carries a Lucas 'Silver King' oil lamp.*

CENTRE: *A James roadster, 1930. This high-quality roadster has a two-speed epicyclic gear incorporated in the front chainwheel, which, together with the driving chain and rear sprocket, is completely enclosed in the gearcase. The brake is an unusual type of centre pull-rod. The end of the rear mudguard was probably painted white in the later 1930s in obedience to a new law. This machine was sold to Mr Fred Clark. He kept it as his 'best' bicycle, using his 1907 BSA (above) for day-to-day riding. It is fitted with a Miller twin-cell battery lamp.*

BOTTOM: *A Chater-Lea gentlemen's bicycle, 1922, made at Letchworth, Hertfordshire. This bicycle was notable for the unusually narrow diameter of the tubing used to build it. The three-pin chainwheel has the initials CL prominently featured in its design. This is a short-wheelbase machine with flat handlebars for touring, a fixed wheel, Pelissier side-pull brakes and a Joseph Lucas oil lamp. Chater-Lea were established in 1890 but at first supplied complete sets of fittings, often made to meausre, to selected local dealers, who would then assemble the machines, supplying wheels and other accessories to the customer's requirements. Shortly after the end of the First World War they began to produce complete bicycles of equally high quality.*

TOP: *A New Hudson ladies' bicycle of 1911. By this date the basic design for a ladies' machine was standardised. It has a curved down tube frame and is fitted with roller lever brakes. The dress cords strung from the rear mudguard to the axle are to prevent the rider's dress getting tangled in the rear wheel. The gearcase is a typical fabric one with celluloid panels fitted around a metal frame. It is fitted with the 'Search Light', an embossed oil lamp decorated especially for ladies' machines.*

CENTRE: *A Golden Sunbeam, 1935, made by John Marston and Company of Wolverhampton. In 1919 the company was acquired by Nobel Industries Limited, which in 1927 was absorbed by ICI, who continued producing the Sunbeam. The frame with its double crossbar is typical of those for tall riders for whom the normal seat pillar and handlebar adjustment were not sufficient. It has internally expanding hub brakes, which gave sure braking under all conditions. Unlike the rim-brake they were fully protected from wet and dirt but, being more expensive, are found only on the best quality machines. Other equipment includes a Sturmey-Archer three-speed gear, oil-tight chain case, a Lucas dynamo lighting set and a rear carrier for luggage.*

BOTTOM: *A Sunbeam ladies' bicycle, 1943. The Sunbeam was manufactured from 1943 by BSA of Birmingham. This machine is fitted with the 'Little Oil Bath Gearcase' developed from the original Harrison Carter enclosed gearcase. In addition to being flawlessly oil-tight it sealed the entire transmission system against dust, mud and water without any noticeable increase in weight and was skilfully built around the frame. Some other makers developed an inferior version which clipped on to the frame. The bicycle has centre-pull roller lever brakes and the rear wheel contains a BSA three-speed hub. A GEC twin-cell battery lamp is carried.*

15

TOP: *A Raleigh Racer, 1892. In the 1890s there was great demand for Raleigh bicycles because of their many racing victories. This success was mainly due to Frank Bowden, who had founded the company in 1888 in Raleigh Street, Nottingham. During 1892 2300 prizes were won by riders on Raleigh machines, notably A. A. Zimmermann. This small-framed racing cycle, with its upward-sloping top frame tube, ball-bearing steering head, plunger front brake and beaded-edge pneumatic tyres, is typical of the period. It has a Lucas 'Silver King' oil lamp.*

UPPER CENTRE: *Ariel Quintette, 1898. This five-seater was built for the Dunlop pacing teams of the 1890s. In 1898 Dunlop had a staff of fifty racers all in regular training, wearing racing suits with 'Dunlop' on the back and with supporting medical staff and mechanics. It has been illegal to use these machines on the open road since 1897; the power produced by the five riders, the strength needed for steering and the lack of brakes made them lethal. The lower frame uses triple tubes to give the necessary strength and rigidity.*

LOWER CENTRE: *The Raleigh low gravity carrier bicycle of 1935 was built with a small front wheel to allow a large basket frame and stand to be fitted. They were used extensively by tradesmen for local deliveries until well into the 1950s. This machine belonged to a lady in Old Harlow, Essex, who used it to carry hay to feed the horses in her stables.*

BOTTOM: *A Gundle carrier, 1950. Unlike the Raleigh carrier this machine is a more straightforward modification of a standard diamond-framed safety. A support for the basket has been fitted above the standard-size front wheel and there is no drop stand. This machine cannot carry as much as the low-gravity machine and with its higher centre of gravity would be much less stable when loaded.*

TOP: *A touring tandem with a Watsonian child's side-car, 1938. Most touring machines by this time were of the 'double-gents' design which gave a much more rigid machine than the 'lady back' or open-frame type. They were ridden by both sexes, the lady wearing a divided skirt, plus-fours or tailored shorts. This machine has drop handlebars and cable-operated hub brakes. Many tandems fitted with child's side-cars remained in use until the 1950s.*

UPPER CENTRE: *A Rudge-Whitworth with a Rann trailer attachment, 1938. This standard Rudge-Whitworth roadster has been converted into a tandem using a trailer attachment made by F. M. Grubb Limited of Wimbledon, a firm which normally specialised in lightweight machines. This trailer could also be fitted to tandems so that parents could have their child riding with them. The bicycle has a Miller twin-cell battery lamp.*

LOWER CENTRE AND BOTTOM: *Sunbeam ladies' and gentlemen's bicycles, 1957. The last Sunbeam bicycles were built by BSA between 1943 and 1957. From 1954 they were produced with the BSA Hublite dynamo in the front-wheel hub and a unique quick-release BSA three-speed hub gear in the rear wheel. One major problem with the oil-tight gearcase was removing the rear wheel, which would normally involve partly dismantling the gearcase before disconnecting the chain. The ladies' machine (lower) demonstrates the last of many Sunbeam innovations, the quick-release rear wheel which left chain and gearcase intact. In 1957 Raleigh took over BSA and brought to an end seventy years of manufacture of the finest-quality bicycle in the world.*

17

earliest days but the development of the safety with pneumatic tyres dramatically reduced race times. By 1893 the record for 100 miles (161 km) was five and a half hours and in the following year the distance between Land's End and John O'Groats was covered in just under three days, six hours. Even ladies' racing was gradually accepted, although sixteen-year-old Tessie Reynolds caused a great uproar in 1893 when she raced from London to Brighton in eight and a half hours. Higher speeds called for lighter machines and steel was replaced by aluminium and other alloys during the twentieth century. The present day use of titanium has reduced the weight of some machines to a mere 10 pounds (4.5 kg) and experiments with carbon fibre in the early 1980s have reduced weights still further.

The early problems and continued popularity of cycle racing is shown by the history of the Tour de France, still the

TOP: *Coventry Eagle ladies' sports bicycle, 1929. This model is fitted with semi-drop handlebars with roller-lever brakes, which on later models were replaced with the caliper type for lightness, and a Joseph Lucas 'King of the Road' oil lamp. The frame, of high-quality steel tubing, is made with a straight tube, from the steering head to a point on the seat tube, almost parallel to the lower frame tube. The wheels have rims of the Westwood pattern and are fitted with 26 by 1³⁄₈ inch (660 by 35 mm) wired pneumatic tyres.*

CENTRE: *This 1938 Claud Butler lightweight, the 'Recognized King of Lightweights', is built with very light high-grade steel tubing and well designed lugs, which make an appealing brazed frame. The rear wheel incorporates a Sturmey-Archer type AM medium ratio three-speed lightweight alloy hub gear. Claud Butler was in the forefront wherever cyclists gathered: at club meets, road events, rallies, world championships and the Olympic Games.*

BOTTOM: *A Royal Enfield ladies' sports bicycle, 1941. Produced during the Second World War, when materials were in short supply, many components that had previously been chromium plated were enamelled. The mudguards are the original celluloid types finished in white so the machine was easier to see in the blackout and the brakes are of the side-pull caliper type.*

world's greatest test of physical and mental endurance in cycling. The race was started by Henri Desgrange in 1903 and covered over 1500 miles (2414 km) in six days. The 1904 race was almost a disaster with riders attacked by hostile crowds and mobbed by over-enthusiastic supporters. Each year the race lengthened, reaching its maximum in 1926 with seventeen stages covering 3565 miles (5737 km). Today the race is shorter but with more stages, allowing greater speeds.

Track racing at places like Crystal Palace and Herne Hill, which originally had a banked wooden surface (it was concreted in 1896), also grew in popularity. The need to pace riders attempting speed records led to the development of cumbersome monsters like the five-seater Ariel Quintette, later superseded by large pacing motor cycles which enabled speeds of over 50 mph (80 km/h) to be attained. During the 1970s and early

TOP: *A Paris Galibier racer, 1948. This machine's unusual frame, which was intended to reduce weight, was designed by Harry Rensch, a talented London frame builder. After building them for established firms he went into partnership to found the Paris Cycle Company to produce the Galibier. They have been used in many races, particularly the Tour de France. The gear is a five-speed Derailleur type which differs only in detail from the first version patented in France about 1909.*

CENTRE: *A Hetchin's lightweight, 1950. The unique feature of these beautiful hand-built machines is the 'Vibrant' or 'Curly' stays of the rear fork triangle. The other Hetchin's features are the hand-cut lug-work of the frame, and the round, finely tapered front forks with the twin-forged fork crown. It is equipped with Chater-Lea lightweight fittings and a five-speed Derailleur gear.*

BOTTOM: *A Viscount Aerospace sport, 1976, made by the Trusty Manufacturing Company Limited of Potters Bar, Hertfordshire. When first introduced these bicycles were described as 'the nearest thing to man-powered flight'. The light and immensely strong frame tubes of 4130 chrome-molybdenum steel are joined with low temperature hand-brazed lugless joints; the complete frame weighs a mere 3¾ pounds (1.7 kg). The bottom bracket and wheel hubs are fitted with sealed ball races. It is equipped with a ten-speed gear and brakes operated by dual levers, so that they can be applied instantly from any position on the handlebars.*

TOP: *A Mike Burrows track bicycle, 1980, a beautifully designed low-profile, hand-built machine. To obtain the very short wheelbase twin seat tubes are fitted to allow the rear wheel to be mounted further forward, and the frame is shortened by mounting the handlebars on the long extension. To give a low profile, it is fitted with front wheels of 24 inches (610 mm) diameter and 27 inches (686 mm) at the rear while retaining flat handlebars.*

CENTRE: *A Kalkhoff professional racing bicycle, 1982, made at Colppenburg, West Germany. Designed and built for high performance, this machine is equipped with Dura-Ace lightweight alloy brakes and a twelve-speed gear. The quick release wheels are fitted with tubular tyres stuck on to sprint rims and pumped up to pressures in excess of 70 pounds per square inch (4920 g/sq cm). The large 1 inch (25 mm) bore pump fitted is essential for inflating such tyres.*

BOTTOM: *A Raleigh cyclo-cross bicycle, 1983. This specially built Raleigh Ilkeston 753 cyclo-cross bicycle was ridden by Kevin Smith in 1983 in the world's toughest cyclo-cross, the Three Peaks International in Yorkshire, and the Smirnoff International at Harlow. It is equipped with a six-speed Derailleur gear. The wheels are fitted with high-pressure tubular tyres on sprint rims, the rear tyre having a knobbly tread to give better grip on rough ground.*

1980s highly specialised, streamlined models were built for the sole purpose of making new speed records.

The greatest variety of shapes were those created for the various delivery bicycles which developed during the early years of the twentieth century from the carrier tricycles which first appeared in the late 1880s. The first one known was made about 1911 and those still in use remain basically unchanged today. The majority had baskets of varying sizes mounted at the front. To accommodate the larger baskets a very clumsy looking frame with drop stand was mounted above a small front wheel. Below the crossbar, or between the two crossbars on the larger machines, a metal plate carried the name of the firm, often painted to order by the makers.

Even before the development of the safety and the bicycle boom of the 1890s some army officers like Lord Wolseley predicted that the bicycle would play a part in warfare because of the extra mobility it could give to infantry. Development of folding machines began and the first British folding bicycle was used

ABOVE: *A BSA military folding bicycle, 1915. Basically a standard 28 inch (710 mm) wheeled safety machine, it was heavy and awkward even when folded. The vertical tube linking the crossbar and bottom tube has a rod inside it with a thread at the base, which screws tightly into the lower part of the frame. When the screw is tightened, the tapered top of the rod is forced into the crossbar to hold the frame rigid. It carries a Lucas oil lamp.*

BELOW: *A Paratroop folder bicycle, 1940, made by BSA. The frame used small-diameter twin parallel tubes for lightness, a similar idea to that used on the earlier Dursley Pedersen. The use of curved tubes gave a frame of unusual form. For folding, the upper and lower arms of the frame are hinged and simply secured by wing nuts. After the war, when many machines were sold off, the simple bar pedals shown on this machine were replaced with standard rubber ones. The bicycles gave infantry greater mobility on the ground and their silent running was especially useful in night operations. Its light is a twin-cell battery lamp with an adjustable black-out attachment.*

during the South African War of 1899 to 1902. By 1914 the British army had fourteen cycle battalions totalling 7000 men. During the First World War bicycles were used on the Western Front and units saw service as coastguards and in India. Because of their speed cycle companies were used mainly to rush reinforcements to threatened sections of the line. Development continued and during the Second World War a lighter BSA folding machine was used by paratroops as well as for general movement. When folded several machines could be fitted easily into gliders and landing craft, leaving room for the troops. After 1945, increasing mechanisation led to the phasing out of the bicycle.

The Quadrant shaft-drive bicycle, 1899, an attempt to replace the chain drive of the safety bicycle with shaft and bevel gears. Other models such as the Rover, Columbia and Belgian FN used gears with teeth, but this machine is fitted with the unique Lloyd and Priest cross-roller drive, patented in 1897. The covers over the gears have been removed to display the gear mechanism. It is fitted with a Power and Hanmer 'Dictator' oil lamp.

ADVANCES AND FAILURES

Bicycle makers have always tried to improve and modify their machines in the search for greater efficiency and higher sales and to produce machines for special purposes. As the safety bicycle developed in the years before John Boyd Dunlop perfected his pneumatic tyre, manufacturers were faced with the problem of providing a smoother, more comfortable ride. One solution was to incorporate a spring system in the frame as in the Whippet of 1885 but development in that field soon halted, not to be taken up again until Moulton re-introduced the idea of front and rear fork suspension in the 1960s.

The efficiency and smoothness of the drive to the rear wheel was greatly improved when the roller-type driving chain still used today replaced the earlier block chain. This did not stop the experiments and several more complicated forms which had theoretical advantages were tried. These included the Simpson lever chain whose form was supposed to give increased leverage. When raced against roller-chain machines it proved to be no better and by 1900 this and other experimental chains were no longer used. Experiments were also made with other forms of transmission. In 1882, the shaft drive with bevel gears was first applied to a tricycle and by the later 1890s several shaft-drive machines like the Quadrant existed. Despite its neat appearance and lack of clothing traps this form of propulsion never caught on as it was no more efficient than the roller-chain drive.

The quest for greater lightness for racing machines led to the use of a range of quite unsuitable materials like wood and bamboo. They gave the necessary lightness but lacked strength, especially at the joints. The most successful and unorthodox of the early lightweights and also the first of the 'de luxe' machines was the Mikael Pedersen named after, and patented by, its inventor in 1893. The prototype's triangulated frame of light duplicated tubes weighed only 14 pounds (6.6 kg) and was stronger than the standard diamond frame, while the ham-

TOP: *A Hadley bicycle with a Simpson lever chain, 1896, made by Begbie and Twentyman of Kentish Town, London. This cycle has a unique sprung seat and the lever chain invented by William Speirs Simpson of London and patented in 1895. The deep triangular side plates have two sets of engaging surfaces, the inner one meshed with the teeth on the front chainwheel and the outer one running on the bearing surface of the double back cog. This was supposed to give greater speed with less effort. However, when raced on a track it was no better than conventional roller-chain machines.*

CENTRE: *A wooden-framed racer of unknown make, 1898. This example is very rare and was in a very decayed state when found. During restoration all the wooden parts had to be renewed. The 28 inch (710 mm) wheels with their unusual wooden rims were obtained from Holland and are fitted with wired pneumatic tyres. Wood provided the necessary lightness but not the strength needed and steel reinforcing pieces had to be fitted at all the angles and joints.*

BOTTOM: *A Mikael Pedersen made in Dursley, Gloucestershire, in 1894. In the early 1890s Mikael Pedersen invented a new type of hammock saddle. As it could be fitted to existing machines only after expensive modifications he rethought the design of the bicycle and the result was the Pedersen triangular frame. His first machine, built in 1893, had a wooden frame. This example uses thin steel tubes and with a weight of only 20 pounds (9.1 kg) was the first successful lightweight machine. It is fitted with a 'Perfecta Nova' acetylene lamp.*

TOP: *A 1906 Dursley Pedersen. This hand-built machine was developed from the Mikael Pedersen and was renowned for its quality, strength, rigidity and lightness. It was a high-quality but expensive gentlemen's machine. The early Pedersen three-speed hub gear, patented in 1902, was probably added at a later date. The frame's tubing is so light and thin that it would not stand brazing and had to be soft-soldered. Many people doubted the strength of the machine because of this but the fears were unfounded and many remained in use for half a century or more.*

CENTRE: *A Triumph Recumbent, 1935. Triumph produced only six of these machines, which were based on a Swedish idea, but they were never popular because of the difficulty of balancing. This is a semi-recumbent design and because of the position of the rider a steering wheel had to be fitted, as conventional handlebars would foul the rider's legs when cornering. This machine did little to lessen wind resistance compared with some of the true recumbents of the 1930s.*

BOTTOM: *The Elswick Scoo-Ped, 1958. One of the first small-wheeled bicycles to be produced after the Second World War, this Elswick-Hopper machine had a Sturmey-Archer three-speed gear, hub brakes, battery lighting and a unique fibre-glass body with legshields which gave some degree of weather protection. Only a few were manufactured and they never became popular, mainly because of their weight and the difficulty of gaining access to the moving parts for repairs.*

mock saddle was both light and comfortable. The later production models weighed about twice as much. The type had considerable popularity and remained in production in varying forms until about 1920, of the machines being ridden for over fifty years. The ladies' model is regarded as the most scientifically constructed ladies' machine ever produced, combining a full triangular and open frame giving the necessary strength and stiffness.

At first glance the steering wheel and seat of the Triumph Recumbent seem to be based on the motor car. Wheel steering was essential here because handlebars would have fouled the rider's legs when cornering. The motor cycle and motor scooter influenced cycle design, although the British cyclist was spared the sight of machines with dummy petrol tanks that appeared in America, and the Elswick was quite neatly disguised as a motor scooter.

Experiments were also made with small size machines which were claimed to be more convenient for town use. These were generally satisfactory for short distances, but the upright position of the rider made riding against the wind or up hills very difficult. 1982 saw the appearance of the first all-plastic bicycle, the Itera. Despite the maker's claims, it failed to attract interest and soon ceased production. The entertainment world, especially the circus, has produced a considerable number of freak machines with one or more wheels. Very little is known about the making of such machines as they were generally produced by individual craftsmen to special order.

There have been two particular developments during the twentieth century which deserve closer examination, the recumbent and the small-wheeled machine. The latter has been the only development which has come near to displacing the traditional diamond-framed safety.

The recumbent or horizontal bicycle first appeared in the early 1930s and is a classic example of a machine ideal in theory but with disadvantages in practice. The first of these machines, the Velocar, was introduced in France in 1933 and its unconventional design and riding position enabled it to capture several world

ABOVE LEFT: *A replica of the Velocino, built by John Collins, Curator of Mark Hall Cycle Museum, in 1982. Although by 1935 the Velocino had become popular in Italy, it was, according to a report in 'Cycling' magazine in 1931, first made in Belgium. The awkward position of the handlebars made steering very difficult and the small degree of movement possible gave the machine a very large turning circle. This replica is fitted with a Lucas twin-cell battery lamp.*

ABOVE RIGHT: *The Cyclo-Ratio recumbent, made by the Cycle Gear Company of Birmingham in 1935. During the 1930s many experimental machines tried to overcome the problem of wind resistance. Here the rider and pedals are in a nearly horizontal instead of vertical line. The rider could exert greater thrust on the pedals and these machines proved very fast in speed trials and could climb well, thanks partly to their excellent Cyclo four-speed Derailleur gear. The chain has 138 links and was difficult to tension correctly. The machine never became popular in Britain; many cyclists did not wish to be seen on such an unlikely-looking contraption.*

BELOW: *A 1934 Cyclo-Recumbent, one of the first attempts to reduce wind resistance by lowering the position of the rider. This is obtained by using an elongated version of the traditional diamond-shape frame and small-diameter 20 inch (508 mm) wheels. The machine is fitted with the first type of Cyclo two-speed Derailleur gear.*

ABOVE: *A 1936 Cyclo-Ratio recumbent. This is an improved version of the 1935 model, the major fault of which was its enormous chain. The frame of this machine is built with two bottom bracket assemblies so that transmission is by two chains. This bicycle is also lower and more stable, being fitted with smaller-diameter wheels. It is equipped with an early Cyclo Derailleur three-speed gear.*
BELOW: *An Avatar-type recumbent built by Mr D. M. Scutchings of Weston Turville, Buckinghamshire, in 1983. This is a much improved version of the recumbents of the 1930s. This design is based on the American Avatar 2000 but is fitted with a Sturmey Archer AW three-speed gear, making it suitable for everyday use. The American version has proved to be a very fast machine on the track as it combines a lightweight frame and light-alloy fittings with a twenty-one-speed Derailleur gear.*

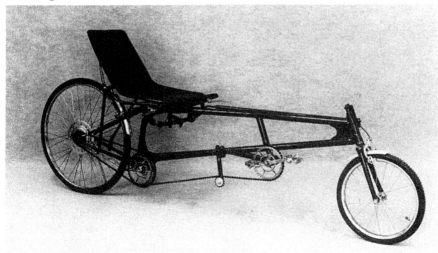

A Mobo pavement bicycle, 1950, probably made in Italy. This type of small children's bicycle first appeared early in the twentieth century and soon became known as the 'fairy cycle'. When this machine was produced they had become known as pavement bicycles and were fitted with stabilisers which were often soon discarded, as has happened with this example. During the 1960s they rapidly became more popular than the traditional chain-driven children's tricycle, which they have now almost replaced.

speed records which were not ratified. The rider lay almost horizontally with the pedals raised above the small front wheel. This position lessened wind resistance and enabled the rider to use both his back and leg muscles to power the machine. This position enabled high speeds to be reached for short periods but after that the rider tired quickly and was liable to considerable thigh fatigue. Standard touring models with dynamo lighting set, chain guard and Derailleur gear were available for a time. These machines never caught on as road vehicles.

Since the Second World War the re-cumbent has reappeared on several occasions and during the early 1980s several designs appeared in the United States including the Vector, Wilkie and Avatar. They are only used for short distance speed trials and when equipped with sophisticated streamlining two, three and four-wheeled versions have reached unpaced speeds well in excess of 60 mph (97km/h). For the narrow purpose for which they have now been developed, these machines are ideally suited.

On the other hand the small-wheeled machine has presented a serious challenge to the standard machine. Small-wheeled machines have been made for

The Moulton Stowaway, 1962. This was the first completely new cycle design since the 1890s. The designer, Alex Moulton, combined the cross frame and small wheels, to which was added a sprung rubber suspension system. This enabled tyres to be pumped up hard, giving a smooth and easy ride. Despite the small wheels the gearing made it as fast as a conventional machine. The first prototype was made in 1959 and was followed by a variety of designs, including this one, the frame of which takes apart (right) for storage, in, for example, the boot of a car.

TOP: *The Moulton Speed Six, 1965. This was the first true light-weight speed machine in the Moulton range, weighing 28 pounds (12.7 kg) when fitted with carrier and mudguards. The front and rear suspension were stiffened to improve its track qualities and it was probably the first mass-produced bicycle in Britain to be fitted with a six-speed Derailleur gear. The carrier and support could be detached when racing. When used for pursuit racing in Canada against standard track machines with carefully matched riders the Moultons invariably won.*

CENTRE: *The Raleigh RSW 16 mark I, 1965. When this bicycle was launched in 1965, it was Raleigh's long awaited answer to the Moulton. The initials 'RSW 16' stood for 'Raleigh Small Wheels 16 inch'. The wheels are fitted with 2 inch (51 mm) diameter low pressure tyres, which were claimed to be as comfortable as the narrow high pressure tyres used on the Moulton. The wide tyre had a greater rolling resistance but looked comfortable to the uninformed potential buyer.*

BOTTOM: *The Moulton 'Mini', 1968. This machine was a ⅞ scale version of the original Moulton and was first introduced in 1966. It was manufactured after the take-over of Moulton by Raleigh in 1967. For the next few years a variety were marketed, most remaining in production until 1974. Depending on the fittings in which it was equipped, it could be used by riders from about six years of age to adults 5 feet 8 inches (1.75 m) tall. Many of the later Raleigh models dispensed with the front suspension.*

TOP: *A Triang 'Junior 1970', made under licence from Raleigh by Lines Brothers of Merton, London. This was the first 14 inch (356 mm) diameter wheel version of the Moulton and was intended as a children's machine for six to ten year olds. The rear suspension was the normal type while the front was semi-sprung, with only the front end of the frame sprung and not the handlebars.*

CENTRE: *A Moulton mark III, 1970. This machine had a completely new triangulated rear suspension assembly and the rear carrier was detachable. The length of both wheelbase and seat tubes was reduced and the three-speed Sturmey-Archer gear incorporated a cable-operated hub brake. Only one version was produced but it became the basis of a variety of high performance conversions. One modified prototype, the 'Marathon', was ridden from England to Australia in 1970.*

BOTTOM: *The Chopper mark I, 1970. Developed from the Raleigh RSW 16, this bicycle was an immediate success when launched in 1970 and a great quantity were sold. Produced for eight to fourteen year olds, it is fitted with a 16 inch (406 mm) ribbed front tyre and a 20 inch (508 mm) knobbly rear tyre for sure road grip. The saddle is a high-back polo type with chrome roll bar.*

The Raleigh BMX, 1984. BMX stands for 'Bicycle Moto Cross' and is another development of the small-wheel bicycle. The BMX machines are for off-road riding and stunts. They therefore have to be built with a very sturdy frame, so the manufacturers have reverted to the use of the well proved traditional diamond frame. The wheels are made of injection-moulded plastic.

children throughout much of the bicycle's history but it was not until the 1930s that the small wheel first appeared for road use, fitted to several of the oddities previously described. Amongst these individual machines, about most of which little is known, was one with 16 inch (406 mm) wheels and an open frame of quite modern appearance photographed on the streets of Paris and reproduced in the German magazine *Signal* sometime between 1940 and 1944.

The appearance of the Moulton towards the end of 1962 marked the arrival of a small-wheeled machine well able to compete with the standard safety. In 1957 Alex Moulton began investigating how the bicycle could be further developed. The small wheel had considerable technical advantages over larger ones and enabled him to produce a machine with a low centre of gravity, ideal for large amounts of luggage. Small wheels inevitably give a bumpy ride thus making some form of suspension essential. Balloon tyres were rejected because they would reduce the speed of the machine and after much experimentation suspension units were fitted at front and rear. Raleigh Industries were invited to manu-

facture the Moulton but eventually refused so Alex Moulton established Moulton Bicycles Limited, which in a short time became the second largest bicycle-frame maker in Britain.

Other cycle makers suffered a considerable loss of sales and investigated other versions and in 1965 Raleigh produced the RSW 16. Unable to use a Moulton-type suspension system, they adopted balloon tyres, making it a much cheaper machine. This and its variants finally obliged Alex Moulton to sell out to Raleigh in 1967. Raleigh continued to produce Moultons and in 1970 launched their RSW mark III, Moulton mark III and the outlandish Chopper, which took most of the limelight. Production of the Raleigh Moulton finally ceased in 1974, but in 1983 Alex Moulton produced a newly designed de luxe Moulton. From the late 1970s increasing numbers of cheap foreign small-wheelers appeared and Raleigh responded with the BMX, which first appeared in 1983. Although ideal for children, the poor riding qualities of these machines deter serious cyclists, with whom the large-wheeled diamond-framed safety, first produced in 1885, remains supreme.

FURTHER READING

Alderson, F. *Bicycling, A History.* David and Charles, 1972.
Bowden, K. *Cycle Racing.* Temple Press, 1958.
Calif, R. *The World on Wheels.* Rosemount, 1983.
Caunter, C. F. *The History and Development of Cycles.* The Science Museum, 1972.
Caunter, C. F. *Cycles, Descriptive Catalogue.* The Science Museum, 1958.
Clayton, N. *Early Bicycles.* Shire Publications, 1986.
Crowley, T. E. *Discovering Old Bicycles.* Shire Publications, 1973.
Dobson, Peter. *The Corgi Book of Bicycles and Bicycling.* Corgi Books, 1985.
Hallard, Tony. *The Moulton Bicycle.* Undated.
Lightwood, J. T. *The Romance of the Cyclists Touring Club.* C. T. C. London, 1928.
Ray, Allan J. *Cycling: Lands End to John-o-Groats.* Pelham Books, 1971.
Ritchie, Andrew. *King of the Road.* Wildwood House, 1975.
Watson, R., and Grey, M. *The Penguin Book of the Bicycle.* Penguin Books, no date.
Williamson, G. *Wheels Within Wheels (The Starleys of Coventry).* Bles, 1966.
Woodforde, J. *The Story of the Bicycle.* Routledge and Kegan Paul, 1970.

PLACES TO VISIT

Boston Guildhall, South Street, Boston, Lincolnshire PE21 6HT.
Telephone: 01205 365954. Website: www.bostonguildhall.co.uk
Coventry Transport Museum, Millennium Place, Hales Street, Coventry CV1 1JD.
Telephone: 024 7623 4270. Website: www.transport-museum.com
Horsham Museum, 9 The Causeway, Horsham, West Sussex RH12 1HE.
Telephone: 01403 254959. Website: www.horshammuseum.org
Museum of Harlow (formerly the Mark Hall Cycle Museum), Muskham Road, Harlow,
Essex CM20 2LF. Telephone: 01279 454959. Website: www.harlow.gov.uk
National Cycle Collection, The Automobile Palace, Temple Street, Llandrindod Wells,
Powys LD1 5DL. Telephone: 01597 825531. Website: www.cyclemuseum.org.uk
National Motor Museum, John Montagu Building, Beaulieu, Brockenhurst, Hampshire SO4
7ZN. Telephone: 01590 612345. Website: www.beaulieu.co.uk
National Museums Liverpool, William Brown Street, Liverpool L3 8EN.
Telephone: 0151 478 4393. Website: www.liverpoolmuseums.org.uk
Riverside Museum Project, Museum of Transport, 1 Bunhouse Road, Glasgow G3 8DP.
Telephone: 0141 2872720. Website: www.glasgowmuseums.com
The Science Museum, Exhibition Road, South Kensington, London SW7 2DD.
Telephone: 0870 870 4868. Website: www.sciencemuseum.org.uk
The Shuttleworth Collection, Shuttleworth (Old Warden), Aerodrome, Biggleswade,
Bedfordshire SG18 9EP. Telephone: 01767 627927. Website: www.shuttleworth.org
Thinktank Birmingham Science Museum, Millennium Point, Curzon Street, Birmingham
B4 7XG. Telephone: 0121 202 2222. Website: www.thinktank.ac
Streetlife Museum, High Street, Hull HU1 1PS. Website: www.hullcc.gov.uk
Ulster Folk and Transport Museum, Cultra, Holywood BT18 0EU.
Telephone: 028 9042848. Website: www.nmni.com/uftm

Printed and bound by CPI Group (UK) Ltd, Croydon, CR0 4YY

11/10/2024

01043558-0008